The Synchronicity of Numbers

A Doorway into Spirituality

Imannie Walshe

Copyright © 2021 By Imannie Walshe

All rights reserved. The content of this book may not be reproduced, duplicated, or transmitted without direct written permission from the author or the publisher. Under no circumstances will any blame or legal responsibility be held against the publisher, or author, for any damages, reparation, or monetary loss due to the information contained within this book. Either directly or indirectly. You are responsible for your own choices, actions, and results.

Disclaimer Notice:

Please note the information contained within this document is for educational and entertainment purposes only. By reading this document, the reader agrees that under no circumstances is the author responsible for any losses, direct or indirect, which are incurred as a result of the use of the information contained within this document.

ISBN: 9798516809835

Dedication

This book is dedicated to my daughter and the entire human family of planet earth.

TABLE OF CONTENTS

Introduction vii

I. What Is Spirituality? 1

II. The Difference Between Spirituality & Religion............7

III. What Is New Thought & Is It the Same a Spirituality 15

IV. What is the Universe...19

V. How the Universe Communicates Through Spirituality 24

VI. A Brief History of Numbers...36

VII. How Numbers Create the Universe 39

VIII. The Synchronicity of Numbers & Messages............42

IX. The Synchronicity of Numbers in Spirituality 45

X. Developing An Awareness to Numbers 63

Conclusion 66

References 69

About The Author..71

INTRODUCTION

The world we live in is full of numbers, and there is relevance in each and every one of them. Each number has a specific meaning that goes beyond a simple quantity, according to Numerology, the divine science of numbers. This frequency is identical to the ones that angels and other heavenly beings resonant with. When our guardian angels want to get our attention, they will frequently send us a series of numbers that will repeat themselves in our experience. The majority of the time, these numbers work through meaningful coincidence or synchronicity. Synchronicity is a term coined by CG Jung, a psychologist who believed that our minds are linked to a universal whole called the Collective Unconscious. Synchronicities are uncanny coincidences that have a lot of meaning. These occurrences feel like a nudge from an unseen guiding force, which we interpret as divine or angelic messages.

You've undoubtedly heard people describe their faith as "spiritual but not religious," and you've wondered what that really means. While the majority of people use the terms "religion" and "spirituality" interchangeably, they are two distinct concepts. So, what's the difference between spirituality and religion? I'll describe both and clarify how they vary in this book. Also, there's an explanation why you're seeing the same numbers over and over. The universe is trying to tell you something, whether it's the time on a clock or the speed on a speedometer. This book will transport you to the origins of the mystery of time. Although it can seem to many as a coincidence, reflecting on these hidden meanings will change everything when it comes to the world and moving forward. This book will teach you what the universe is and how it communicates with us through spirituality. The synchronicity of numbers and how we can apply this numbers in our daily lives.

I
What Is Spirituality?

Spirituality is a general term that refers to a belief in something other than oneself. It can include religious traditions centered on the belief in a higher force, but it can also include a holistic belief in an individual's relationship to others and the rest of the universe. It implies that something greater binds all things to one another and to the world as a whole. It also suggests that there is life after death and aims to provide answers to questions about the nature of life, how humans are related to one another, universal truths, and other mysteries of human existence.

Spirituality provides a perspective on life that says there is more to life than what people see and feel on a sensory and physical basis. Many people have found spirituality and religious practice to be a source of relaxation and stress relief. Although people seek God in a variety of ways and follow a variety of paths to communicate their faith, research has shown that those who are more religious or spiritual and use their spirituality to cope with life's struggles reap numerous health and well-being benefits.

SIGNS OF SPIRITUALITY

Spirituality isn't limited to a particular direction or set of beliefs. Spirituality and the advantages of a spiritual experience can be experienced in a variety of ways. For some, this may include faith in a higher power or adherence to a particular religious tradition. Others may feel a sense of interconnectedness with the rest of humanity and nature, or feel a sense of belonging to a higher state. Asking serious questions about subjects such as loneliness and what happens after death are some manifestations of spirituality. Developing stronger bonds with others, having compassion and empathy for humanity is a wonderful way to express spirituality, too. Feelings of interconnectedness, awe and wonder, beyond material possessions or other tangible incentives is another example of spirituality. All ways can lead to eventual contentment or happiness.

Spirituality is not experienced or expressed in the same way by all. Some people pursue spiritual experiences in all aspects of their lives, while others are more likely to have these feelings in particular circumstances or places. Some people are more likely to have spiritual encounters in churches or other religious temples, while others are more likely to have these emotions while out in nature. Spirituality can take many forms, ranging from religious rituals to more secular approaches. Buddhism, Christianity, Hinduism, Humanism, Islam, Judaism, New Age Spirituality, and Sikhism are some of the main types of spirituality. It's worth noting that there are a variety of other spiritual traditions around the world, including traditional, African, or Indigenous spiritual practices. Spiritual activities like these can be especially relevant for groups of people who have been affected by colonialism.

Spirituality can be pursued for a variety of purposes, including, but not limited to, seeking meaning and purpose. Spirituality can help people find answers to existential questions like "what is the meaning of life?" and "what is the purpose of my life?" To deal with stress, depression, and anxiety, among other things. When dealing with life's struggles, spiritual experiences can be beneficial. Spirituality is used by people who want to rekindle their faith and optimism. Spirituality may aid in the development of a more positive attitude toward life. Even, to find a sense of belonging and assistance. Being a member of an organized religion or community, which is common in spiritual practices, may be a valuable source of social support.

Although individual spiritual beliefs are a matter of faith, research has shown that spirituality and spiritual practice can have a number of benefits. The findings would come as no surprise to someone who has sought solace in their religious or moral beliefs, but they are remarkable in that they show in a scientific manner that these practices do help a large number of people. Spirituality and its effect on physical and mental wellbeing have yielded a slew of promising results.

RELIGION & SPIRITUALITY

Religion and spirituality have been shown in studies to help people deal with the consequences of daily stress. According to one study, daily spiritual encounters helped older adults better deal with negative emotions while also increasing positive emotions. According to study, people who have an intrinsic religious orientation have less physiological reactivity to stress than people who have an extrinsic religious orientation, regardless of gender. Many who were intrinsically oriented devoted their lives to God or a "higher force," while those who were extrinsically oriented used faith for external purposes such as making friends or improving their social status in the group.

This, along with other findings, suggests that staying involved in a spiritual group may have positive and long-term benefits. This participation, combined with the appreciation that can accompany spirituality, can act as a stress reliever and has been related to improved physical health. Dedication to God or a "higher force" resulted in reduced stress reactivity, improved feelings of well-being, and, eventually, a reduced fear of death. People who are at ease with using spirituality as a stress coping strategy should be assured that there is more proof that this is a good thing for them. Prayer is effective for both young and old. Better health, greater psychological well-being, less depression, less hypertension, and less stress have all been attributed to prayer and faith, particularly during tough times. Combined they promote more optimistic emotions and a better ability to deal with tension.

Exploring your spiritual side may help you improve your well-being, whether you're rediscovering a forgotten spiritual path, strengthening your commitment to an already well-established one, or finding a new source of spiritual fulfillment. Spirituality is a deeply personal experience, and each person's spiritual journey is different. However, research indicates that certain spiritual stress reduction techniques, regardless of faith, have proven to be beneficial to many people.

Paying attention to how you feel will help you explore spirituality. Accepting spirituality entails accepting both the positive and the negative aspects of human nature, which could lead to putting others first to release focus on one's self. Spirituality includes things like opening your heart, having empathy, and helping others. Relax and think about it. Spend 10 to 15 minutes in the morning doing some kind of meditation. Gratitude should be practiced. Start a gratitude journal and write down what you are thankful for every day. This will serve as a helpful reminder of what matters most to you and gives you the most joy. Try mindfulness; by practicing mindfulness, you can become more conscious of and grateful for the present moment. Mindfulness allows you to be less judgmental of yourself and others, and to concentrate more on the current moment rather than the past or future.

Spiritual bypassing is one of the possible pitfalls of spirituality. This is characterized by a proclivity to use spirituality as a means of avoiding or avoiding difficulties, feelings, or disputes. Instead of apologizing for an emotional wound you've caused somebody else, you might actually excuse it by implying that "everything happens for a reason" or that the other person should "reflect on the better." Spirituality will enrich your life and provide a variety of benefits, but you must be careful not to let spiritual values contribute to traps like dogmatism or an excuse to disregard the needs of others.

II
The Difference Between Spirituality & Religion

Religion is a term used to describe an organization that has a collection of formal activities and a structured belief system that is shared by and among its members. Their transcendental values, which are often passed down from generation to generation, are founded on either a formally documented creed or existing cultural practices. In any case, each faith has professionals who serve in leadership roles and represent the formal aspects of the organization. The leaders often perform rituals related to the religion's core values, which serve as the basis for how one should live one's life.

Religion serves as an extremely supportive social network because it is a community or group of people who share the same beliefs. It has practical consequences for daily activity in terms of adhering to the group's values, traditions, and activities. Members of a religion frequently adhere to specific dress codes, moral codes, and acts that are dictated by a supernatural being (especially religious leaders). A religious individual is dedicated to adhering to his or her religion's rules. He or she follows religious rituals and customs, such as attending Sunday church services for Christians, keeping the Sabbath for Jews, and fasting during Ramadan for Muslims. Religion is a term used to describe an organization that has a collection of formal activities and a structured belief system that is shared by and among its members.

Spirituality, on the other hand, is concerned with one's inner self and soul. Being spiritual entails adhering to one's own collection of values and practices while still looking for the meaning of life. Through personal research and self-reflection, each person's concept of spirituality will change over time, adjusting to their unique experiences. Spirituality's aim is to recognize one's intrinsic power to overcome all of life's obstacles and to communicate with something bigger than oneself. Spirituality is a universal, personalized experience that is special to each individual. A spiritual experience may be defined as sacred or transcendent, or simply as a true sense of aliveness and connectedness, or as pure gratitude. You may be spiritual when belonging to a religious community, and vice versa.

Indeed, certain people's spirituality may be directly associated with a church, while others may have their own personal relationship with a higher power. Others look for meaning in their lives through their relationships with nature or art. In the controversy between faith and spirituality, neither side is necessarily good or evil. Spirituality is essentially undefinable, whereas religion has a fixed, concrete code of ethics. Both of these ways of believing in something, on the other hand, help people live peaceful, fulfilling lives. A spiritual person is distinguished from a religious person by the following characteristics.

Individual vs. Group Beliefs: Religion is a community of people brought together by their shared faith or convictions about the divine, while spirituality is a single experience of the divine. Religion is meant to help people develop their character. It influences one's values, attitudes, and behaviors by emphasizing the importance of following laws. This brings people of different religions together because they have similar characteristics and outlooks on life. Spirituality, on the other hand, focuses more on the individual soul of each person.

There are many religions in the world, but each preaching that their narrative is the right story. Spiritual people, on the other hand, pick and choose what they believe from every religion and incorporate these truths to create their own collection of beliefs. As a result, the answer to the question "Can you be spiritual and religious?" is "yes—if your true convictions align with a particular religion." Many spiritual people, on the other hand, assume that everyone's true reality is the same, regardless of their differences.

Spirituality is more concerned with the content of the message being delivered than with the discrepancies in the original story's data. While Christians may pay close attention to Moses' story and the Ten Commandments, a spiritual person may take away wider messages from the story, such as being thankful for what one has or always being truthful.

While everyone's spiritual views are different, spirituality also reinforces the notion that everyone is the same. There are no borders, races, or ethnic divisions in the world. Love is at the heart of everyone's being, and humanity as a whole is one. Regardless of ideological differences, all humans share the same consciousness, which manifests itself in different ways. Since everyone's nature is the same, any distinctions are merely cosmetic. This realization inspires people to love one another and encourage one another on their journey to enlightenment.

Progression of Beliefs

A spiritual person's collection of values grows as they learn more about their own faith through personal research. In faith, on the other hand, the belief system is normally predetermined. The religious group's recognized authorities reinforce or express changes in the philosophies that members of the religious group should follow.

Religion requires people to be true to their faith in this way, keeping them fixed on scriptures or what religious leaders teach them. Religion also discourages people from following their intuition and often encourages them to obey and follow orders without question; of course, there are exceptions to all rules. However, people who do practice religion are expected to be true to their beliefs. In most cases, the value system is predetermined.

Spirituality, on the other hand, encourages people to listen to their inner voice continuously and use it as a guide to live a good life. Spirituality encourages people to break away from conformity and submission in order to forge their own course in life. As a result, someone who practices spirituality can change their beliefs frequently throughout their lives. Nothing is set in stone, so what one considers "right" or "wrong" today can change with new life experiences down the road.

Spirituality Has No Set of Rules

Rather than pursuing an ideology or set of principles, a spiritual individual often discovers their own truths when developing their spirituality. The experience is usually intimate and takes place in secret, allowing people to trust their instincts and act in their own best interests. Although some spiritual traditions provide theories and practices to aid people on their spiritual paths, they are not intended to be taken as a system of beliefs. Rather, they're presented as resources for spiritual seekers to use in order to aid their spiritual growth.

Religious people embrace the truth as established by their faith, and these truths are often written and shared with others. The promise of punishment or reward for keeping beliefs or performing rituals is common in religion. The reward for spirituality, on the other hand, is simply one's own inner peace. The fear of being punished for one's conduct is a common motivator in religion. Many who practice religion are concerned with the long-term repercussions of their behavior. People frequently believe that if they do not spend their lives in accordance with their religions, they will be damned.

Spirituality, on the other hand, helps people to concentrate their attention on good things and to live purely on the basis of love. Spiritual people do what they believe to be right, regardless of the consequences. While spirituality does not threaten punishment for living a life that violates a set of rules, it frequently discusses karma. This is a cause-and-effect principle in which one's actions or intentions have a direct impact on one's future. This suggests that someone who lives with good intentions and does good deeds will be happy in the future, but someone who lives with poor intents may suffer in the future.

Consider whose laws you obey when you ask yourself, "Am I religious or spiritual?" Do you depend on a particular church's rituals to get you through times when your faith is waning? Can you put your faith in an organization to arrange it for you, or do you pick and choose what you believe? When trying to figure out whether you're religious or spiritual, these are crucial questions to think about.

Origination of Beliefs

As a result of their own personal experiences, spiritual people often learn and develop in their beliefs (often through trial and error). This encourages people to discover their own truths without being limited by their level of knowledge. Religious people, on the other hand, learn or grow their beliefs through parables or the experiences of their religious founders. People are inspired to forge their own path through spirituality, enabling them to embark on their own spiritual journey. When it comes to spirituality, the possibilities for self-discovery are endless. Spirituality helps people to follow their instincts and listen to their hearts.

Spirituality is simply the recognition that each person is more than a physical body; they are a soul with infinite potential. Every person is essentially a mini-god, a unique and powerful manifestation of existence with a specific role to play in the universe. Spiritual people learn and grow their beliefs based on their own experiences, while religious people learn and develop their beliefs based on the experiences of their religious founders.

So, how do you tell the difference between a religious and a spiritual person? A religious person seeks advice from others, while a spiritual person relies on themselves to do the right thing. Spiritual people have the ability to adapt to new circumstances and develop new skills during their lives. When it comes to their beliefs, they are open-minded. Between religion and spirituality, there is an important difference between believing and being.

Religion is concerned with the substance of believers' ideals as well as how those values are expressed in their everyday lives. Spirituality, is more concerned with the process of being aware of one's own inner self. In spirituality, one might ask, "Where do I find meaning in life, and what are my true values?" "What is the truth?" and "What is right and wrong?" are common religious questions. Individual experiences are where the two intersect, influencing how people think, feel, and behave.

III
What Is New Thought & Is It the Same as Spirituality

The New Thought movement was one of the most well-known movements that arose in the nineteenth century to assist people in gaining a greater understanding of spiritual mysteries by using the influence of their minds. The word "New Thinking" referred to the idea that one's thoughts could unlock the keys to living a better life without being bound by religious doctrines or dogmas. Higher Thought (not to be confused with 'New Age') is a spiritual and intellectual movement focused on theological and metaphysical beliefs as well as theories from mental science. During the nineteenth century, the New Thought Movement mostly began in the United States. New Thought, despite its name, is not a new movement; much of the theories, values, and core teachings that inspired it are rooted in ancient wisdom. Almost all religious and spiritual practices throughout the world have incorporated these ancient teachings.

New Thought is not a faith, a denomination, or a spirituality in and of itself. Since 'Modern Thought' is essentially ancient wisdom articulated in a new way, it differs from spirituality, although there are many religious groups or movements with New Thought origins. In his classic book, Varieties of Religious Experience, William James referred to New Thought as "The Religion of Healthy-Mindedness." As a spiritual movement, New Thought aided in the direction of a variety of social shifts in the nineteenth, twentieth, and twenty-first centuries. New Thought had a direct influence on the mid-nineteenth-century "Mental Sciences" movement, which eventually became known as the New Thought Movement. The mental-healing movement was a reaction to outdated beliefs and practices, especially old-school medical practice and theology. The new thought movement is spiritually oriented, but it is not synonymous with spirituality.

The New Thought Movement is a religiously oriented or metaphysical understanding of New Thought beliefs that originated in the United States in the 1800s. In England, the word "Higher Thought" was often used to describe the campaign. The New Thought Movement arose from the reaction of conscious thinking people revolting against rigid religious dogma in the mid-nineteenth century, who desired a blending of science and philosophical values. The roots of the New Thought movement are usually attributed to Phineas Parkhurst Quimby's mental healing methods and Ralph Waldo Emerson's transcendentalism.

New Thought's theories were often expressed in books, magazines, and leaflets during its early stages. It wasn't designed to be some kind of organized religion or religious organization. The New Thought movement's theories were widely disseminated around the world, primarily through seminars, books, and journals. Today, the movement is made up of a loosely affiliated community of religious groups, writers, thinkers, and individuals who believe in metaphysics, positive thinking, the law of attraction, healing, life force, imaginative visualization, and personal control.

Religious Science (or, The Science of Mind), Unity Church (or, Unity Society of Practical Christianity), and the Church of Divine Science are the three main divisions and religious groups within the New Thought movement. There are a few smaller organizations as well, the majority of which are part of the International New Thought Alliance. Nonjudgmental, open congregations, New Thought churches make everyone feel unique, welcome, and respected. While the New Thought Movement is often thought of as a single movement, the thoughts and ideas of different groups may differ. However, there are certain shared values that run across all of the groups and serve as the foundation for the New Thought Movement's overall philosophy.

The "New Age," or the New Age Movement, is often confused with New Thought, or the New Thought Movement. New Thought is simply a theory that has been around for a long time. It is, in essence, the reality thread that runs through all of the world's great spiritual practices. The spiritual teachings and doctrines that influenced much of New Thought have centuries-old origins. The new thought's values are universal.

New Thought adherents believe that "God" or "Infinite Intelligence" is "supreme, universal, and everlasting," that divinity dwells inside each individual, that all people are spiritual beings, that "the highest spiritual principle is loving one another unconditionally." (KJV)

God is regarded as Eternal and the Cause and Source of All in New Thought, as well as in mainline Christianity; moreover, in mainline doctrine, God's goodness is often contrasted with any independently existing evil (personified as satin or the devil), or in a dualistic context (separation from God). New Thought believes that God is present within, yet infinitely exceeding, the manifest universe, and that the spiritual universe is organized by metaphysical laws that can be activated by the use of spiritual practices to consciously create life experience, similar to how the physical universe can be described by observed physical laws.

IV
What is the Universe

The universe is everything, it encompasses all of space, as well as all of the matter and energy contained within it. It also involves the passage of time and, of course, you. Earth and the Moon, as well as the other planets and their hundreds of stars, are all part of the universe. The planets, along with asteroids and comets, orbit the Sun. In the Milky Way galaxy, the Sun is one of hundreds of billions of stars, and several of those stars have their own planets, known as exoplanets. The Milky Way is one of billions of galaxies in the visible universe, all of which are believed to have supermassive black holes at their cores, including our own. The universe includes all the stars of all the galaxies, as well as everything else that astronomers can't see. It's all, in a nutshell.

While the universe appears to be a mysterious place, it is not far away. Outer space is just 62 miles (100 kilometers) away from wherever you are right now. If you're sleeping, enjoying lunch, or dozing off in class, outer space is only a few hundred miles above your head, day or night. It's also beneath you. The unforgiving vacuum and radiation of outer space is about 8,000 miles (12,800 kilometers) below your feet, on the opposite side of the Earth.

In reality, you're currently floating in space. As if Earth is different from the rest of the world, humans say "out in space" as if it is there and we are here. But, like the other planets, Earth is a world that exists in space and is a part of the universe. Things exist here by chance, and the atmosphere near the surface of this particular planet is hospitable to life as we know it. In the universe, Earth is a small, delicate outlier. The entire universe is a hostile and merciless world for humans and other living beings on our planet.

It appears that the universe is 13.8 billion years old. The number was calculated by combining the ages of the oldest stars with the rate at which the universe expands. They have calculated the expansion by looking at the Doppler change of light from galaxies that are almost all moving away from us and each other. The galaxies are moving apart at a faster rate as they get farther apart.

The motion of galaxies should be slowed by gravity, but now they're speeding up, and scientists aren't sure why.

The galaxies will be so far away in the future that their light will not be detectable from Earth. To put it another way, last Saturday, matter, energy, and everything in the universe (including space itself) were all more compact than they are today. Any time in the past, last year, a million years ago, a billion years ago, can be defined in the same way. However, the past does not last indefinitely.

Scientists discovered that before galaxies developed or stars started fusing hydrogen into helium, objects were so close together and hot that atoms couldn't form and photons had nowhere to go by calculating the speed of galaxies and their distances from us. All was in the same place a little farther back in time. Or, more accurately, the whole universe (not just the matter in it) was contained in a single location.

However, don't get too excited about a mission to visit the location where the universe began, since no one will go to the Big Bang's birthplace. It's not as if the world started out as a dark, hollow space and then exploded, releasing all matter. There was no such thing as the cosmos. There was no such thing as space. Since time is a part of the universe, it did not exist. Time started with a big bang, as well. As the universe evolved over time, space itself expanded from a single point to the vast cosmos.

How Our View of the Universe Changed Over Time

The way people think about the world, how it functions, and how big it is has evolved over time. Humans had little or no knowledge of the cosmos for countless lifetimes. Instead, our forefathers relied on legend to justify the roots of all. The myths represent human concerns, dreams, expectations, or fears, rather than the essence of truth, since they were created by our forefathers.

Humans started to apply mathematics, literature, and new investigative principles to the quest for information many centuries ago. Those ideas, as well as experimental tools, were refined over time, gradually revealing clues about the existence of the universe. When people started actively studying the existence of things just a few hundred years ago, the term "scientist" didn't even exist (researchers were referred to as "natural philosophers" for a time). Since then, our understanding of the cosmos has advanced in leaps and bounds. Astronomers first discovered galaxies outside our own about a century ago, and humans have only been sending missions to other planets for about a half-century.

Space probes have traveled to the far reaches of the solar system and returned the first close-up images of the four giant outermost planets and their countless moons; rovers have wheeled around the surface of Mars for the first time; humans have built a permanently crewed, Earth-orbiting space station; and the first large space telescopes have delivered jaw-dropping views of more than a billion stars. Astronomers discovered thousands of planets around other stars, observed gravitational waves for the first time, and captured the first glimpse of a black hole in the early twenty-first century.

Humans continue to reveal the mysteries of the universe through ever-advancing technology and intelligence, as well as an abundance of creativity. This quest is aided by new experiences and motivated ideas, which also spring from it. And though there are billions upon billions of other stars in the galaxy, we have yet to send a space probe there. Even in our own solar system, humans haven't visited any of the planets. In other words, the vast majority of the universe that can be understood is still unknown.

The world is about 14 billion years old; our solar system is 4.6 billion years old, life on Earth has been around for 3.8 billion years, and humans have only been around for a few hundred thousand. To put it another way, the world has been about 56,000 times longer than our human race. By that standard, almost all that has ever happened occurred prior to the arrival of humans. So, of course, we have a lot of questions because, in a cosmic sense, we've just recently arrived.

V
How the Universe Communicates Through Spirituality

The world is attempting to communicate with us. It's attempting to pique our interest. It's attempting to lead us, link with us, and arouse us. There are plenty of indicators, but are we looking for them? Are we paying attention? Do we really know what we're looking for? How many times have you wished for a sign indicating the path you should follow in life? How many times have you needed a response but weren't sure how to get it? How much do you feel alone and lonely in your life? Life becomes a lot less complicated when you're tuned in to how the world is interacting with you, and you start to feel a lot more aligned and linked.

We do, however, have physical limitations because we exist in a physical environment. Since these physical limitations prevent Spirit from tapping you on the shoulder and saying, "Yes, I think you should take that work," we must pay attention to the more subtle and imaginative ways Spirit communicates with us. The trick is to SLOW DOWN, as in live life at a slower pace in general, and to BE AS PRESENT AS POSSIBLE so that you can tune into whatever sign appears in your life. You'll never notice when Spirit is trying to get your attention if you're lost in your head worrying, overthinking, overanalyzing, rushing and being busy, judging and being frustrated. You must be present, available, and ready to receive information.

Every sign I receive is a miracle to me because it represents a direct link with Source. When I see a sign, I know Source is right there with me, talking and directing me, and I am aware that I am surrounded by pure Love. This has really been a miraculous experience. The thing is, these miracles aren't what you would expect or as depicted in film. There are no angels falling from the sky (at least not actually and physically), and no face in the sky talking to you and offering advice.

Universal Spirit

The signs are less ethereal and have become part of the processes and habits of daily life. The signals are subtle and indirect, and they can be quite gentle and quiet, but they can also be very loud and clear. Universal Spirit isn't just out there; it's in all and everyone. It's all around us, and as a result, it's entwined in everyday life. This means you could get a sign when you're in the shower, doing the dishes, arguing with your girlfriend, or stuck in traffic. A symbol may not have to be glamorous for it to exist, but seeing one in these less-than-glamorous moments would undoubtedly make them more so.

Also, each sign that comes your way has a meaning that is special to you and your situation, and you see signs when and where you're supposed to see them. As a result, signs are not a one-size-fits-all proposition, and you must approach them intuitively and from a more conscious standpoint. It's critical that you put your ego and fears aside and allow your instincts to lead you. With that out of the way, let's look at some of the ways Universal Spirit interacts with us!

Humans are the vehicles in which the Universal Spirit communicates with us. I'm not sure how it happens, but the world has a way of using people to deliver those messages. This can happen when you encounter a random person who says or does something that sparks something inside you or causes a deep knowing inside you, when you receive a text or phone call from another, when you hear a message on the radio, or when you overhear a conversation.

Obviously, not all you hear is a warning, but it IS a sign when you are looking for a particular response and it is given to you by someone who is randomly asking you something that is related to your question. You'll instinctively know that this is far too random to be anything other than a cosmic warning. The important thing now is to really DO IT. For example, if you're debating whether to keep or leave your current job because you've been offered a promotion, and you're panicking and freaking out because you have to make a decision quickly, and then you get a text message from one of your coworkers saying, "I hate this job, they totally screwed me, they lied about my raise," I'd say that was a red flag.

The Synchronicity Appearances

A synchronicity is a series of events that appear to be associated, coincidental, or serendipitous, but are not connected by something. For example, you see a child with a red balloon on the way to work, then you arrive at work and learn that someone gave your coworker red balloons for her birthday, and then you see another red balloon on an "open house" sign as you drive home. Is it all a fluke? No way, I'm afraid. Synchronicities are one of the most important ways in which the world interacts with you, and the more awaken you are, the more you can notice them. And sometimes it's not even necessary to decipher the synchronicity, it's just a matter of understanding that something far BIGGER is going on behind the scenes. Sometimes, it's something much more profound than we can ever imagine. Synchronicities are a warning that you're on the right track because they're a direct link to Spirit.

Numbers

Seeing a particular pattern or sequence of numbers, such as on license plates, or encountering the same number or number pattern, such as in an address or phone number, or seeing the time 11:11, 2:22, 3:33, 12:34, and so on, can all have hidden meanings. If you see repeated numbers like 1111, 333, 444, 555, and so on, that's a hint from the universe that you're on the right track. If you look at what the numbers mean in numerology, each number series may have more complex meanings. You will also see random numbers that have unique meaning for you, such as your favorite, lucky, or birthday number.

Songs

Have you ever had a song pop into your mind out of nowhere? Or did you hear a certain song on the radio? Or hearing a particular collection of lyrics on someone else's radio? Or hearing someone singing a song as they pass you by? When these things happen at random and you have a deep emotional attachment to the song you hear, it's probably worth looking into more. Pay attention to the lyrics you hear because they can contain a message for you. View the context of the sign from that perspective whether it's a familiar song, such as your wedding song, a song that reminds you of a holiday, or a song that you liked in high school.

Random Thought

You may have an idea that comes to you out of nowhere. It might be a random thought, a solution to a dilemma, or something you're motivated to do. Pay close attention to this and consider what it might mean for you. For example, if the thought "I wonder how my brother is doing" occurs to you, it is likely a sign that you should call your brother, as he is likely in need of your love and support right now.

Word/s or Physical Signs

A sign may be a physical sign with words or pictures on it, such as a road sign, billboard, or Instagram post that is attempting to communicate with you. If you've been in a really dark position recently and see a post on Instagram that says "the darkest hour is just before dawn," I'd take that as a sign to trust that things will get better soon. If you see a billboard with a palm tree on it that says "it's time for a holiday" right after debating whether or not to take a break from work, I would take it as a sign to do so. If you see a road sign that says "slow down" and it resonates with you for some reason, it may be a sign that you need to slow down and be more present in your life. Again, there are a plethora of ways to interpret signs; the trick is to trust your instincts and not overthink things.

Dreams

Dreams will provide us with messages, instructions, and answers. Have you ever had a vivid dream that felt eerily real? Who did you run into? What did they have to say? What went wrong? How did you feel at the time? What new knowledge did you gain? Do you find yourself having the same dreams over and over? What are they for, exactly? Are you constantly seeing the same guy, going to the same places, and getting the same kinds of experiences? Investigate the "meaning of dreams" and see what it might imply. Do you have frequent nightmares? What are they for, exactly? Perhaps it stems from a deeper insecurity that needs to be addressed. Do you have recurring dreams about a certain period in your life or being with people from your past? What does this imply for you personally? Is there a lingering question that needs to be resolved?

Symbols from Nature

Colors, trees, plants, feathers, rocks, and crystals are all associated with specific meanings. Pay attention to what you see, what you come into touch with, what you're attracted to, and what keeps reappearing in your life. Hawks, for example, represent increased spiritual awareness and a time to fly high; the color red, on the other hand, represents peace, prosperity, and good fortune; and a feather falling on your lap is a "hello" from the cosmos and a sign that your angels are close by.

Deep Feeling or Knowing

This is a direct indication from the universe when you feel like "this is right," "something is wrong," or "every part of me wants to do this." Also, when you hear a message or read something in a book that profoundly resonates with you and you don't understand why, but it feels right and real, that's your inner truth awakening. There is a profound truth within us that is linked to Universal Truth, and hearing this Truth awakens something deep within you that makes you feel strongly drawn to it.

Objects

When you discover something, you've been searching for, when objects fall off shelves or out of cabinets, or when you come across an item that activates a memory, supports a new concept, or answers a query, the world will interact with you through objects. Someone offers you a book about financial independence, for example, just as you plan to get out of debt. Or you're out walking in the woods and come across a random baby toy on the ground. Or you come across a photo of your sister, whom you haven't spoken to in years. Alternatively, a random book falls from a bookstore's bookshelf. Spirit is trying to tell you something in all of these ways; it's up to you to find out what it is.

Technological Malfunctions or Failures

Maybe your internet goes down, or there's no service or Wi-Fi, or your text message won't send, or your machine shuts down, or your movie won't load, or your TV won't turn on. These symbols can be viewed in a variety of ways. If you have no service but want to send a text, it may be the universe's way of reminding you that now is not the time to work and that you should be enjoying the moment. When a text message refuses to send, it could be an indication that you did not send it or that you should reconsider/reword what you sent. If your television isn't working, it may be an indication that you're watching too much television.

Issues in Your Body

Physical pain, symptoms, ailment, and illness are all signs that something is wrong with you from the world. Your lifestyle and activities may be contributing to an issue, such as back pain from excessive sitting. This could indicate that you need to alter your lifestyle to become more involved. Illnesses, on the other hand, may have supernatural causes that manifest as physical symptoms. For example, if you're "itching to get out" of a situation or relationship you don't like, you could get a rash. To get to the bottom of what's actually causing a symptom, it's a good idea to look at both lifestyle AND spiritual and emotional triggers. When you cure the emotional component, the physical symptom always goes away on its own.

Setbacks / Roadblocks / Detours / Delays

When your flight is delayed, your interview is canceled, you become ill, and so on, life could be attempting to nudge you in a different direction. Since these things are normally beyond our control, it's best not to fight or oppose them and instead embrace them as signs. It's possible that the world is attempting to reposition you on the right track. It may also be an indication that your energy is low vibrational and causing these setbacks, in which case you can change your energy and attitude to match with a higher vibration. "Be grateful for the closed doors, detours, and roadblocks that life throws at you. They guard you against roads and locations that aren't meant for you." Said Suzanne Heyn, a writer whom lives in New York City.

Everything is Falling Apart

There are moments in life where everything appears to be falling apart, and just when you think things can't get any worse, they do. Your car breaks down, your roof leaks, you lose your work, your child gets into a fight, your husband or wife dumps you, or someone you care about passes away unexpectedly. When everything in our lives falls apart, it's always a sign that we need to learn to let go and withdraw from the world of structure. It's a chance to reawaken. It's also a great chance to reconsider and reevaluate how we've been living our lives, see what needs to improve, and start again in a way that's true to who we are.

Your Life Flowing Together & Aligning

This is the polar opposite of anything falling apart, and you're feeling motivated, healthy, and in the zone, with opportunities popping up left and right, your desires manifesting, you're feeling linked and aligned, and things just seem to work out. This is a lovely indication that you're on the right track and consistent with your reality.

While in Meditation

Great wisdom is unlocked when you are in a state of presence and go inside to communicate with yourself. This allows Spirit to connect with you in a more direct way, allowing you to obtain signs, instructions, and answers. Everyone around you is doing something you're thinking about or instinctively know you should do. These are all simple signs from the universe that say, "Yes's A word of caution: if any of the above symptoms appear repeatedly in a short period of time, or if they appear during your life, you should pay special attention to them.

Persistence and repetition are sure signs that the world is trying to communicate with you. It's important not to overanalyze, overthink, or attempt to make sense of these signals logically as they appear to you. It stops being a warning when you over-investigate it. The universe's wisdom is expressed in great simplicity; it is just our minds and egos that try to figure out why, when, and when. So always go with your gut instinct!

The argument is the Universal Spirit will ALWAYS be on your side. It's here to help you and answer your questions. Its aim is to protect and help you. It exists to remind you that you are not alone and that you DON'T HAVE TO WALK THIS PATH IN FEAR, DISCONNECTION, OR CONFUSION. The universe is aware of your needs until you are aware of them too. All you have to do now is watch for the signs and TRUST that you will be led and encouraged along the way. All has a divine order and timing to it; believe it. Finally, remember that YOU ARE Universal Spirit. It is the ESSENCE of your personality. So, in the end, it's your own self that leads you to where you're meant to be.

VI
A Brief History of Numbers

Since the dawn of time, humanity has had a love-hate relationship with numbers. Scratches on the bones of people who lived 30,000 years ago may have represented lunar phases. The ancient Babylonians reported the motions of the planets as numbers and used them to predict eclipses and other astronomical events. The ancient Egyptian priesthood used numbers to forecast the Nile's flooding. Pythagoreanism was a Greek cult that believed numbers were the foundation of the world, which operated on numerical harmony.

Pythagorean theories were a combination of foresight (numerical characteristics of musical sounds) and mysticism (3 is male, 4 is female, and 10 is the most perfect number). For supernatural purposes, numbers were paired with names: the biblical "number of the beast," 666, is most likely an example of this practice. More recently, kooks have found the secrets of the cosmos in the dimensions of Giza's Great Pyramid, a phenomenon so widespread that it has its own name: pyramidology. Millions of otherwise reasonable people are afraid of the number 13, to the point that hotels avoid using it on their floors, airlines don't have a row 13, and Formula 1 racing cars' numbers skip from 12 to 14 so that, for example, 22 cars can be numbered from 1 to 23.

Despite countless theories to the contrary, learned tomes have been written about the meaning of such stalwarts as the golden number (1.618034), which occurs in flowering plants and modern architecture but not in the shell of the nautilus or ancient Greek architecture. Many religions, as well as organizations like Freemasonry, have sacred numbers; Wolfgang Amadeus Mozart's music, especially the Magic Flute (1791), contains many deliberate references to Masonic numerology.

Numbers are old, significant, and potent. The Pythagoreans were one of the first groups to popularize the notion that numbers are more than just abstract representations, but also have spiritual meaning in the 6th century BC. "Number is the lord of forms and thoughts, and the cause of gods and daemons," Pythagoras is said to have said. Pythagoras is responsible for the numerology scheme, which assigns various meanings to numbers. But numerology is just one side of the story: numbers have been assigned special meanings throughout history, religion, and mythology. For centuries, numbers have held significance in cultures all over the world, from the Far East to our own western heritage.

VII
How Numbers Create the Universe

All in the universe, including humans, is part of a mathematical system, according to Tegmark. He claims that all matter is made up of particles with properties like charge and spin, but that these properties are strictly mathematical. Space, on the other hand, has properties like dimensions but is essentially a mathematical framework. "If you embrace the notion that both space and everything in space have no properties at all but mathematical properties," Tegmark said in a talk at The Bell House on Jan. 15, "the idea that everything is mathematical begins to sound a little bit less crazy." His talk was inspired by his book "Our Mathematical Universe: My Search for Reality's Ultimate Nature" (Knopf, 2014).

"If my theory is incorrect, science is doomed," Tegmark said. "There's nothing we can't, in theory, comprehend," he said if the world were really mathematics. The concept is based on the observation that nature is full of patterns, such as the Fibonacci sequence, which is a series of numbers in which each number is the sum of the two numbers before it. The distance between each petal and the next in the flowering of an artichoke, for example, corresponds to the ratio of the numbers in the series. The nonliving universe follows a statistical pattern as well. A baseball would follow an approximately parabolic trajectory if thrown into the air. Planets and other astronomical objects travel in elliptical orbits.

"There's an elegant simplicity and elegance in nature uncovered by mathematical patterns and shapes that our minds have been able to find out," said Tegmark, who has famous equations framed in his living room. Because of the statistical structure of the universe, scientists may theoretically predict any physics observation or measurement. Mathematics predicted the presence of Neptune, radio waves, and the Higgs boson particle, which is thought to describe how other particles get their mass, according to Tegmark. Some claim that math is nothing more than a method developed by scientists to understand the natural world. Tegmark, on the other hand, claims that the mathematical framework found in nature proves that math exists outside of the human mind. Can we use math to understand the brain, while we're on the subject of the human mind?

Mathematics of Consciousness

The human brain has been dubbed "the most complex structure in the world" by others. Many of the great leaps in understanding our universe have been made possible by the human mind. Tegmark believes that one day, scientists will be able to use math to explain consciousness. "It's likely that consciousness is how knowledge feels when it's being processed in certain, very complicated ways," Tegmark said. He noted that many major advances in physics have included the unification of previously disparate concepts such as energy and matter, space and time, and electricity and magnetism. He believes that the mind, which is the sensation of a conscious self, will eventually be merged with the body, which is made up of moving particles.

But, if the brain is all logic, would it rule out free will because particle motions can be measured using equations? He stated emphatically that this was not the case. To put it another way, if a machine wanted to mimic what a human would do, the computation would take at least as long as the action itself. Some have proposed that free would be described as the inability to foresee what one will do before an event occurs. However, this does not imply that humans are weak. "Humans have the power not only to understand our environment, but also to form and change it," Tegmark said at the end of his talk.

VIII
The Synchronicity of Numbers & Messages

How often have you come across numbers like 11:11, 777, 1010, 4444, 999, and so on? Seeing repeated numbers is a form of synchronicity, which is described as "a meaningful coincidence of two or more events where something other than the likelihood of chance is involved" by psychiatrist Carl Jung. To put it another way, synchronicity is something more than serendipity, which is based on chance and fate. The unconscious and etheric realms are where synchronicity is born. When we have synchronicity, our unconscious mind and Higher Self are sending us signals and messages.

Jung, for example, identified a synchronicity in which he saw a half-man, half-fish sign. After that, he was served fish for lunch. Someone joked about turning another into an "April frog." One of his patients showed him a photograph of a fish in the afternoon. Another individual showed him an embroidery of sea monsters and fishes later that evening. His next patient told him about a dream she had about a fish the next morning.

Synchronicity can be unsettling, especially when certain symbols – in this case, numbers – appear repeatedly. If you keep seeing the same number, it's important and try to figure out what significance it holds for you. The numbers, symbols, and words that you keep hearing have a lot of meaning in your life, and you should pay close attention to them.

The phenomenon of synchronicity has been labeled as "meaningful coincidence," a paranormal occurrence, angel numbers, the cosmos speaking to us, pseudoscience, and so on. In his book Synchronicity: An Acausal Connecting Principle, Swiss Psychologist Carl Jung introduced the idea of synchronicity as a phenomenon of energy with noncausal events (1952). It sparked numerous theoretical debates, and although those debates continue today, people continue to claim to have experienced synchronicity. You wouldn't be reading this because you're going through it yourself or just curious about what the "meaning" of it all is.

People encounter synchronicity numbers almost everywhere, and they are often unexpected. The numbers have been found on shopping receipts, prices for products in stores and on the products themselves, license plates, in video games, in movies or shows, on the time stamp of a YouTube video or a song, the random page number of a book, mile markers on the road, house numbers, billboard advertisements, and several other places. It is entirely up to those who come across these numbers to determine if they are a meaningless coincidence or a troubling warning. Many that assign value to these numbers would be easily labeled as ludicrous, and will strive to avoid doing so in the future, but the more it happens, the harder it will be to ignore.

Synchronicity numbers are closely associated with a form of revolutionary event or spiritual awakening that occurs in one's life, so the meanings of the numbers below would be associated with those events. Those who study numerology, occultism, theology, and other related subjects are often the ones who contribute to the meanings of these numbers.

IX

The Synchronicity of Numbers in Spirituality

A small sample will reveal an enormous variety of symbolic roles that numbers have played in different cultures, religions, and other systems of human thought.

NUMBER 1

The number one is, unsurprisingly, regarded as a symbol of unity. As a result, it often represents God or the cosmos in monotheistic religions. Since number means plurality and 1 is singular, the Pythagoreans did not consider 1 to be a number at all. They regarded it as the source of all numbers, however, since adding many 1s together can produce any other (positive whole) number. In their scheme, even numbers were male and odd numbers were female, but the number 1 was neither; instead, it switched from one to the other. When you add 1 to an even number, it becomes odd, and when you add 1 to an odd number, it becomes even.

When you hear the number one three or four times, you're being told something interesting. This number is all about fresh starts and finding a new place to call home. The more you see this one, the more you'll notice little positive actions in your daily life. You could get some good news, discover something you've been looking for, or even run into an old acquaintance. This number tells us that you're doing something good in the world, and the universe is taking notice.

NUMBER 2

Many simple dualities are represented by the number two: me/you, male/female, yes/no, alive/dead, left/right, yin/yang, and so on. Dualities abound in human perceptions of the universe, owing to our preference for two-valued logic—another duality, true/false. While the Pythagoreans considered 2 to be a female number, other numerological systems considered it to be a male number. 2 is a symbol for man, sex, and evil in Agrippa von Nettesheim's De occulta philosophia (1533; "On the Philosophy of the Occult").

The biblical book of Genesis does not use the formula "and it was good" when referring to the second day of Creation, which may be one explanation why some people equate 2 with evil. Some religions are dualistic, with two gods replacing the monotheistic Deity. For example, in Zoroastrianism, Ahura Mazd (the god of light and goodness) battles Ahriman (the god of evil) (the god of darkness and evil). The number two is often associated with negative connotations, such as duplicity and two-facedness. Since they assumed that mystical forces would bring twins' wishes to fruition, Northwest Coast Indians forced the parents of twins to follow a number of taboos.

When we see the number 222 a lot, it has a lot of influence over our lives. This number indicates that our angels or those who protect us are trying to communicate with us. This may be a vital message foreshadowing anything to come, or just a friendly reminder that they're still around. The more you see this one, the more important it is to pay attention.

NUMBER 3

The number three is a magical and sacred number that appears in numerous folktales (three wishes, three guesses, three little pigs, three bears, three billy goats gruff). The three main gods of ancient Babylon were Anu, Bel (Baal), and Ea, who represented Heaven, Earth, and the Abyss, respectively. Similarly, the Egyptian sun god had three aspects: Khepri (rising), Re (midday), and Atum (evening) (setting). God the Father, God the Son, and God the Holy Spirit are known as the Trinity in Christianity. Plato saw 3 as a metaphor for the triangle, the most basic spatial shape, and believed the universe was made up of triangles.

A paper triangle with a cross in each corner and a prayer in the center was thought to defend against gout as well as a cradle from witches in German folklore. When summoning spirits, three black animals were often sacrificed. A three-colored cat, on the other hand, was a guardian spirit. Three witches appear in William Shakespeare's Macbeth (1606–07), and their spell starts, "Thrice the brindled cat hath mewed," evoking such superstitions. In addition, the smallest magic square has a dimension of 3 in which every row, column, and diagonal sums to 15. The number three is a wonderful number to come across. It demonstrates that as we progress, we are remaining optimistic and radiating energies that the planet needs. As an individual, you are evolving and aligning with your highest type. Take this as a reminder that you're on the road you've always wanted to take in life.

NUMBER 4

The four elements of earth, air, fire, and water, as well as the four seasons, the four points of the compass, and the four phases of the Moon, make up the number four in the universe (new, half-moon waxing, full, half-moon waning). The Four Noble Truths are the essence of Buddhism. The tetractys $1 + 2 + 3 + 4 = 10$ was the root of the most perfect number to the Pythagoreans. The four humours (phlegm, blood, choler, and black bile—hence the adjectives phlegmatic, sanguine, choleric, and melancholic) were believed to exist in medieval times, and the body was bled in different locations to put these humours into harmony.

With four classes of gods (superior, ally, subordinate, and spirit), four forms of animals (creeping, flying, four-legged, and two-legged), and four ages of humans, the number four is fundamental in the Sioux worldview (infant, child, mature, and elderly). Their medicine men told them that they should do it in groups of four. Since 4 is a practical, material number, it has few superstitions attached to it. In China, however, the number four is considered unlucky because she ("four") and shi ("death") sound alike. The Four Horsemen of the Apocalypse wreak havoc on mankind in the biblical Revelation to John.

When you see number 4 many times in a row, you should know that the universe values your efforts. You are tenacious and always ready to go above and above. Although it may seem that no one is paying attention, the forces around you and the light beings who have followed you through the years are proud of you.

NUMBER 5

The first even and odd numbers (2 + 3) add up to 5. (As the sum of the female 2 and the male 3 in the Platonic and Pythagorean traditions, it symbolizes human existence and—in the Platonic and Pythagorean traditions—marriage.) The five regular solids were discovered by the Pythagoreans (tetrahedron, cube, octahedron, dodecahedron, and icosahedron; now known as the Platonic solids). Just four of these were recognized by early Pythagoreanism, so the discovery of the fifth (the dodecahedron, with 12 pentagonal faces) was a bit of a shock. Because of this, 5 was often regarded as exotic and rebellious.

The number 5 was synonymous with the Babylonian goddess Ishtar and her Roman counterpart, Venus, and the five-pointed star, or pentagram, was their symbol. Because of this association with the goddess of love, a knot tied in the shape of the pentagram is known as a lover's knot in England. The number 5 is significant in Manichaeism: the first man had five sons; there are five elements of light (ether, wind, water, light, and fire) and five elements of darkness (ether, wind, water, light, and fire); and there are five elements of darkness (ether, wind, water, light, and fire). There are five sections of the body, as well as five virtues and vices.

The Maya set a fifth point in the center of the four points of the compass, which was significant to them. The five fingers of the human hand, as well as the five extremities of the body, added a sense of mystery to the number 5. (two arms, two legs, head). A human in a circle with outstretched arms and legs approximates the five points of a pentagon, and a pentagram results when each point is joined to its second nearest neighbor. This geometric figure is important in occultism, and it is used in summoning spells to lure a demon or devil, which can then be forced to do the sorcerer's bidding. The assumption that 5 was sacred resulted in the addition of a fifth aspect to the standard four that comprised a human being. The term quintessential comes from this fifth essence, or quintessence.

The number 5 is considered sacred in Islam. The five pillars of Islam are first and foremost: faith declaration (shahdah), prayer (alt), fasting during Ramadan, alms (zakt), and pilgrimage to Mecca (the hajj). Every day, five prayers are said. There are five types of Islamic law and five prophets who give law (Noah, Abraham, Moses, Jesus, and Muhammad).

The secret meaning of this angel number is that you are starting to discover who you are. You are going through changes and positive experiences in your life that can seem to be challenges at the time but will pay off handsomely in the long run. Even though you're up against a lot and may not be prepared for what's ahead, you'll emerge unscathed and strong on the other hand.

NUMBER 6

Six is both the sum $(1 + 2 + 3)$ and the product $(1\ 2\ 3)$ of the first three numbers, thanks to a wonderful series of mathematical coincidences. As a result, it is regarded as "great." A perfect number (excluding itself) equals the sum of its divisors in mathematics, and 6 is the first perfect number in this sense since its divisors are 1, 2, and 3. The number 28 is the next perfect number. There are no known odd perfect numbers, but it hasn't been proven that there aren't any. The number 6 appears six times in Genesis' six days of creation, with God resting on the seventh day. The composition of the Creation is similar to the sum $1 + 2 + 3$: light is created on day 1, heaven and earth appear on days 2 and 3, and all living beings are created on days 4, 5, and 6.

NUMBER 7

The metaphysical 3 plus the material 4 equals 7. Students studied the trivium (grammar, rhetoric, and logic) as well as the quadrivium (music, arithmetic, geometry, and astronomy), a total of seven subjects known as the liberal arts in medieval education. Since there are seven distinct notes in the musical scale—roughly corresponding to the white notes on a piano—Pythagorean interest in mathematical patterns in music gives 7 a privileged position.

The exceedingly harmonious octave is the eighth note up the scale, which is how the name came in. Since it is a prime number, which means it cannot be obtained by multiplying two smaller numbers together, the number 7 is also considered lucky and has a certain mystique. The seven days of the week (Sunday, Moon-day, Tiw's-day, Woden's-day, Thor's-day, Frigg's-day, Saturn-day) are named after various ancient gods and planets. Tiw was a

Norse god of war, similar to Mars in function but etymologically to Zeus, and Frigg was the Old English version of Frea (or Freya), Woden's (= Odin's) wife.

Shakespeare wrote of man's seven ages, a concept that dates back far further. In China, the number seven specifies the stages of a woman's life: a child gets her "milk teeth" at seven months, loses them at seven years, completes puberty at two and a half years, and menopause at seven and a half years. The Moon's phases last about seven days, with 4 7 = 28 days in a month and also in a woman's menstrual cycle. Many societies saw the planets (Sun, Moon, Mercury, Venus, Mars, Jupiter, and Saturn) as "wandering bodies," as opposed to "fixed stars," which stay in the same place in the night sky. According to English scholar Robert Graves, the seven candles of the Jewish menorah that burned in the Tabernacle symbolized the Creation and may be linked to the seven ancient planets.

There were seven roads to heaven and seven celestial cows in ancient Egypt, and Osiris led his father through seven underworld halls. In Christian culture, the seven deadly sins are well-known. The Rosicrucian's fundamental number was seven, which they used as the basis for their text Chymische Hochzeit Christiani Rosenkreutz (1459; Alchemical Wedding of Christian Rosy cross). The number seven was also significant in Mithra's cult, which believed that the soul ascended to paradise via seven planetary spheres. It's possible that the Christian concept of seven layers of purgatory is related.

In folklore, the number seven is often mentioned. Breaking a mirror will bring you bad luck for the next seven years. A cat has seven lives in Iran, not the nine lives in Western mythology. Three and seven are the most popular numbers in the Indian Vedas. Agni, the god of fire, can create seven flames and has seven wives, mothers, or sisters. The sun god's divine chariot is drawn by seven horses. There are seven sections of the earth, seven seasons, and seven celestial fortresses in the Rigveda. The cow has a total of 21 (37) titles. In Hippocratic medicine, the number 7 governs bodily illnesses, with debilitating illnesses lasting 7, 14, or 21 days. In Germany, it was thought that treating pigs with asphodel-laced water for seven days would prevent them from contracting hog cholera. A fever can be healed in Jewish magic by taking seven prickles from seven palm trees, seven chips from seven beams, seven nails from seven bridges, seven ashes from seven ovens, and finally seven hairs from an old dog's beard.

NUMBER 8

Numerologists consider the number 8 to be a lucky number. Any odd number less one has a square that is always a multiple of 8 (for example, 9 1 = 8, 25 1 = 8 3, 49 1 = 8 6), a statement that can be mathematically proven. The gods existed in seven realms plus an eighth world, the fixed stars, according to Babylonian mythology. As a result, the number eight is frequently associated with paradise. According to Muslims, there are seven hells but eight paradises, which represent God's mercy. The eight petals of the lotus, a plant associated with luck in India and a favorite Buddhist symbol, are thought to be lucky in Buddhism.

In China, just as the number 7 determines a woman's life, the number 8 determines a man. A boy gets his milk teeth at the age of eight months, loses them at the age of eight years, reaches puberty at the age of two and a half years, and loses sexual virility at the age of eight and a half years. The Yijing, which describes a divination scheme based on yarrow stalks, has 64 = 8 8 configurations.

NUMBER 9

In comparison to the number 8, the number 9 is often associated with depression or pain. The Ninth Psalm, according to the 16th-century Catholic theologian Peter Bungus, foreshadows the arrival of the Antichrist. The universe in Islamic cosmology is made up of nine spheres—the standard eight of Ptolemy, plus a ninth added around 900 CE by Arab astronomers Thbit ibn Qurrah to describe the equinox precession. 9 appears regularly in Anglo-Saxon cultures.

Nine measures were used to quantify distance in legal situations by the early settlers of Wales; for example, a dog that has bitten someone may be killed if it is nine steps away from its owner's home, and nine people attacking one constituted a legitimate attack. Land ownership in Germany ended after the ninth century, according to

German law. The number 9 appears in a lot of folklore. It is said that a stitch in time saves nine. Cloud nine is the pinnacle of bliss. There are nine lives in a cat. The River Styx, which was used to ferry souls to the underworld in Greek mythology, is said to have nine twists.

NUMBER 10

As previously mentioned, the number ten was the Pythagorean symbol for perfection or completion. Humans have ten fingers and ten toes on each hand and foot. Counting on fingers most likely contributed to the decimal number system, with its symbols 0–9 and position values such as the 7 in 703 being 7 hundred, 7 tens in 173, and 7 units in 507 Round numbers, such as 100 or 1,000, are considered to be powers of ten. However, 10 has no special significance, and any number from 2 onwards can be used as a number base. Computers do, in fact, use base 2, or the binary number system, which uses only the symbols 0 and 1 to represent numbers.

Mathematicians differentiate between "genuine" properties of numbers, which are valid regardless of the notational basis, and "accidental" properties, which exist only because of the notational system—for example, 153 (the number of fish in John's Gospel) is the sum of its digits' cubes, $13 + 53 + 33 = 1 + 125 + 27 = 153$. The occurrences of 10 and its powers are so popular that listing them here is pointless. The Ten Commandments of the Bible, on the other hand, are worthy of mention, particularly because Buddhism has its own set of Ten Commandments—five for monks and five for laypeople.

NUMBER 11

The number 11 has a negative connotation since it is sandwiched between the two auspicious and significant numbers 10 and 12. The number 11 has no relation to the supernatural, according to Bungus, and medieval theology refers to the "11 heads of error." The number 11 is also associated with the zodiac since one of the 12 zodiacal signs is concealed behind the Sun at any given time. Enuma Elish Tiamat, the god of chaos, is aided by 11 monsters in the Babylonian creation myth. The ancient Roman version of a police force consisted of 11 men tasked with apprehending criminals. Several sports have 11-person teams (American football, football [soccer], cricket).

NUMBER 12

The number 12 is closely associated with the heavens—the 12 months, the 12 zodiac signs, and the 12 lunar and solar stations. Twelve main northern stars and twelve main southern stars were known by the ancients. In a day, there are 24 = 2 12 hours, 12 of which are daytime and the other 12 are nighttime. The number 12 is the sum of the numbers of life and good fortune (5 + 7), and it is the product of the religious and secular (3 4). As a result, it possesses a wide range of qualities. It is the number of Christ's disciples in Christianity, and it appears several times in the Bible—

for example, the Twelve Tribes of Israel. Several cultures have used numbers based on 12 (duodecimal); one well-known example is the 12 inches in a foot.

NUMBER 13

Triskaidekaphobes believe the number 13 is unlucky, particularly when it falls on a Friday, a fear heightened by the explosion that nearly destroyed the Apollo 13 lunar spacecraft in 1970. Skeptics point out that, unlike any other manned spacecraft that has crashed, it returned to Earth safely, rendering its crew some of the luckiest people on the planet. While Judas Iscariot was the 13th person to arrive at the Last Supper, the fear of 13 has negative undertones that date back far further, most likely because an extra 13th object spoils the auspicious 12. The Maya and Hebrews considered 13 to be auspicious since there are 13 lunar months in a year (with a slight error). $13 = 10 + 3$ (Commandments and Trinity) in medieval theology, and hence the number had some positive connotations.

NUMBER 14

The number 14 is an even number with attributes similar to those of 7. A period of 14 days is half of the Moon's 28-day cycle, so it takes 14 days (one fortnight, short for fourteen-night) for the Moon to wax from new to full or to wane from full to new. In ancient Egypt Osiris was cut into 14 parts. The number is important in Islam; the Arabic alphabet contains 14 Sun letters and 14 Moon letters. In medieval Germany 14 innocent beings gave legal protection to whomever they accompanied.

NUMBER 15

15 has religious significance since it is the sum of two holy numbers (3 and 5). The goddess Ishtar was served by 15 priests in ancient Nineveh, and the city had 15 gates. The magic constant of the 3 3 magic square is 15, and it was identified with Ishtar in Babylon.

NUMBER 16

Since 16 is the square of 4, it is endowed with positive qualities. The Vedas mention 16-fold incantations, and the Chinese-Indian goddess Pussa has 16 swords, so it was common in ancient India. Nature, according to the Rosicrucian's, is made up of 16 elements.

NUMBER 17

In ancient times, the local deity was offered 17-fold sacrifices in the Urartu region, near Mount Ararat. On the 17th day of the second month, the biblical Flood began and ended on the 17th day of the seventh month. According to Greek mythology, the best day to cut wood for a boat is the 17th of the month. Some Sufis believe that God's most sacred name is made up of 17 letters. Mathematicians considered 17 odds since a regular 17-sided polygon can be built using Euclidean tools such as ruler and compass, a fact discovered by German mathematician Carl Friedrich Gauss when he was 19 years old.

NUMBER 18

Because 18 is twice the number 9, it has some significance when paired with 9. Haldan has 18 sons, and Odin knows 18 things, according to Norse mythology. The number 18 is sacred to the Sufi mystics known as whirling dervishes in the West, and it was customary for guests to carry gifts in multiples of 18. There are 18 books in the Hindu Mahabharata, and there were originally 18 blessings in the Jewish prayer shemone esre (Hebrew: "eighteen").

NUMBER 19

Eclipses of the Sun occur every 19 years on average. The Babylonians thought the 19th day of the month was unlucky because it was 49 days from the start of the previous month (add 30) and, because $49 = 7\ 7$, it was a day of great significance for good or bad. The meaning of the word Wid (Arabic: "One"), an important term for God, in Islamic numerology is 19.

NUMBER 20

The number 20 has no magical meaning, but it is historically significant since it was used in the Mayan number system. To estimate the number of days in the year, the Maya replaced 20 20 = 400 with 20 18 = 360 while counting time. Many old units of measurement include the number 20 (a score)—for example, in pre-decimal British currency, 20 shillings to the pound.

NUMBER 100

Since we use the decimal (base 10) notation for numbers, the number 100 takes on a meaning that it would not have if we used other systems of notation. It's a great number with traces of perfection. The decade (10 years), century (100 years), and millennium (1,000 years) are the three basic units of the Western calendar, with the century being the most common. As a result, the 20th or 21st centuries are used to define a large historical period. A century in cricket is a major accomplishment for a batsman, but being out for 99 runs is a serious disappointment. A half-century (50) is also a positive sign of ability, while falling short at 49 is not. (We would write 49 as 100 if we had seven fingers and counted in base 7, so 49 would presumably be considered an excellent score in such a culture.)

The dollar is divided into 100 cents, and several other currencies (such as the pound sterling and the euro) follow suit. The boiling point of water is 100 degrees Celsius on the Celsius temperature scale. For e.g., the Roman centurion did not always command exactly 100 men when he said "a hundred." Similarly, 101 can also mean "a lot," but it is clearly larger than 100, and its lack of roundness makes it sound more concise, as in the Disney film 101 Dalmatians (1961).

X
Developing An Awareness to Numbers

When it comes to making tough and vital life choices, numbers have proved to be completely reliable. Numbers will help you figure out when it's the best time to start a new business or ask for a raise. It has the power to transform your life by revealing the purpose of your spirit and directing you down a path where you can depend on someone for a fulfilling personal or professional relationship.

It can also help you get closer to opportunities by revealing the factors that lead to positive outcomes. It is important to know the person you will walk through life with inside and out. Numbers will help you understand your partner's interests, goals, mood, trip, and how they intend to arrive at their port of call. They can be compared to road signs, as they alert you about both positive and negative conditions in your life. Numbers may also reveal what makes certain people excel at a job while others fail.

Numbers show your characteristics and assist you in making the best decisions possible in areas such as relationships, wellness, education, marriage, love, and finance. It assists you in setting goals, preparing and guiding your efforts to achieve them, and wriggling your way out of difficult situations. When you struggle to change your life on a regular basis, it means you're missing out on important details. Numerology will help you uncover this information and change your life by letting you know what events are likely to arise in your life and what you can do about them as the person you are.

Have you ever considered how picking the right wedding date will affect your relationship? Numbers will assist you in this endeavor. By analyzing the energies of a land, a neighborhood, or a town, numbers will help you stay in the right position. You can't go back in time to change your past, but you can still make the most of your future, and Numerology is the key to doing so. Why not take a chance? Allow yourself to be on the lucky side of things.

You're supposed to break down the meanings of each number and then merge them if you see a series of synchronicity numbers that appear repeatedly and in the same order. The numbers can also be added together, according to numerologists. The 215 above, for example, will be 2+1+3 = 6. So, the synchronicity is warning you that some kind of perfection is on the way once you learn to trust yourself more (2), be more mindful of how you think (1), and be open to the changes ahead (5).

CONCLUSION

Numbers assist us in discovering the world's hidden significance. They can serve as our personal guide to life, revealing our purpose and calling, as well as our strengths and weaknesses. Numbers can be used to better understand the world and ourselves as individuals, allowing us to gain insight into our intents and personality traits. Using numbers to calculate things like one's life path number, expression number, and heart's desire number, among other things.

Religion is a term used to describe an organization that has a collection of formal activities and a structured belief system. It is a community or group of people who share the same beliefs. Members of a religion frequently adhere to specific dress codes, moral codes, and acts that are dictated by a supernatural being. Being spiritual entails adhering to one's own collection of values and practices while still looking for the meaning of life.

In the controversy between faith and spirituality, both can help people live peaceful, fulfilling lives. New Thought is not a faith, a denomination, or a spirituality in and of itself. The movement arose from the reaction of conscious thinking people revolting against rigid religious dogma. Humans are the vehicles in which the universal spirit communicates with us.

There are many ways to interpret signs; the trick is to trust your instincts and not overthink things. Learn and adopt a religion or spirituality that is best suited for your unique inner spirit; you will know. Trust yourself! If you see a sign that says "slow down" and it resonates with you for some reason, it may be an indication to slow down. Remember that dreams too provide messages, instructions, and answers. Pay attention to what you see daily, what you come into contact with, and what you're attracted to. Start to keep a daily journal. Journal of various signs the universe sends your way through numbers, symbols, words, sounds, or conversations.

We must continue to learn that symbols from nature and numbers are associated with specific meanings that come into our awareness by way of the universe's help. Give respect to this universal assistance that helps each of us along our daily path of life. It is up to us to design how to use the information given to us freely. We may use it to learn more about ourself. We may use it within our religious practices or use it to enhance our practice of spirituality. However, we choose to use it, we must understand that it is ours individually, sent from a source much higher than each of us.

REFERENCES

1. Brown, H.H. (1903) New thought primer, origin, history, and principles of the movement: S lesson in soul culture. Now Folk.

2. Carl G. Jung (1960), Synchronicity: An Acausal Connecting Principle, Princeton University Press, 2012.

3. Diaonis, Persi; Mosteller, Fredrick (1989). "Methods of Studying Coincidences". Journal of the American Statistical Association. American Statistical Association.

4. F. David Peat, 1999, Time, Synchronicity and Evolution.

5. Giles Sparrow (2017) COSMOS: A Field Guide. Quercus Publishing.

6. Larson, M.A. (1985) New Thought, or, A modern religious approach: The philosophy of health, happiness, and prosperity. Philosophical Library.

7. Joy Woodward, A Beginner's Guide to Numerology: Decode Relationships, Maximize Opportunities.

8. Kirby S. Synchronicity: The Art of Coincidence, Choice, and Unlocking Your Mind.

9. Main, Roderick. 2000. "Religion, Science, and Synchronicity." Harvest: Journal for Jungian Studies 46(2).

10. New Thought religious' movement. https://www.britannica.com/event/New-Thought.

11. Ross, T. Spiritual Meaning of Numbers.

12. Trevor Ross, Spiritual Meaning of Numbers Paperback.

13. Woodward, J. (2019), A Beginner's Guide to Numerology: Decode Relationships, Maximize Opportunities.

ABOUT THE AUTHOR

A gifted metaphysician, writer, and spiritual life coach, Imannie Walshe has taught many people worldwide to discover the inner spiritual self. Within this book Imannie teaches the importance of developing an awareness to everyday synchronicities and numbers that can be a doorway into the learning of spirituality. Imannie Walshe expresses within each lecture, that the universe is constantly gifting us unique signs, symbols, and messages.

www.ingramcontent.com/pod-product-compliance
Lightning Source LLC
Chambersburg PA
CBHW070814220526
45466CB00002B/657